One Dimensional Standing Waves in Classical and Quantum Mechanics

Jorge Monreal, Ph.D.

First printing: 2025

Jorge Monreal, Ph.D.
Madison, FL

Chapter 1

Introduction

Reinforcing separation of variables as a mathematical method to transform partial to ordinary differential equations.

In a typical introductory mechanics class, the topic of waves on a string typically follows the periodic motion of a frictionless mass-spring system where the equation of motion is a second order ordinary differential equation (ODE). The most general solution to the ODE is introduced and after application of boundary conditions, as well as initial conditions, a specific solution is found. Students at the level of a calculus-based introductory physics course are not expected to guess at the most general solution since most have not had differential equations yet. However, they do have the mathematical training to be able to apply boundary and initial conditions since they need only make substitutions, use algebra, and be able to calculate derivatives for the initial condition. Examples of physical systems that lead to periodic motion include a mass-spring system, simple pendulums, and an LC circuit

$$\frac{d^2x(t)}{dt^2} + \omega^2 x(t) = 0 \tag{1.1}$$

$$\frac{d^2\theta(t)}{dt^2} + \omega^2 \theta(t) = 0 \tag{1.2}$$

$$\frac{d^2q(t)}{dt^2} + \omega^2 q(t) = 0, \tag{1.3}$$

respectively. All are second order ODEs undriven and undamped [4]. The difference in angular frequency for these systems depends on the natural oscillation parameters for each.

Introductory-level students struggle with the equation of motion for waves on a string because it is described by a second order partial differential equation (PDE). Since most students have not had a PDE course yet, a cursory talk mentions traveling waves arising from the equation, as well as standing waves arising from boundary conditions encountered by incident and reflected waves. Unfortunately, for most students there seems to be no real connection between these standing waves and the equation of motion. It appears that, for the most part, students dismiss the equation of motion as being irrelevant to standing waves.

Here, we show that it is possible to introduce standing waves which arise from boundary and initial conditions to the second order PDEs of motion in a manner appropriate for students who already examined oscillatory systems such as the ones mentioned above. It is hoped that the examples shown here provide introductory physics students in STEM fields with additional insight into the physics and math which describes oscillatory motion.

The second order PDEs case chosen here has two themes in common. First, the equations of motion are described by second order PDEs and through use of separation of variables, the problem is turned into second order ODEs, which have been previously solved. Second, all cases describe standing waves. To show the extent of standing waves across different physical phenomena we offer three cases: One-dimensional mechanical waves on a string, plane electromagnetic waves on perfect conductor, and one dimensional probability waves of a quantum mechanical particle in a box.

Chapter 2

Classical Mechanics: Mechanical Standing Waves on a String

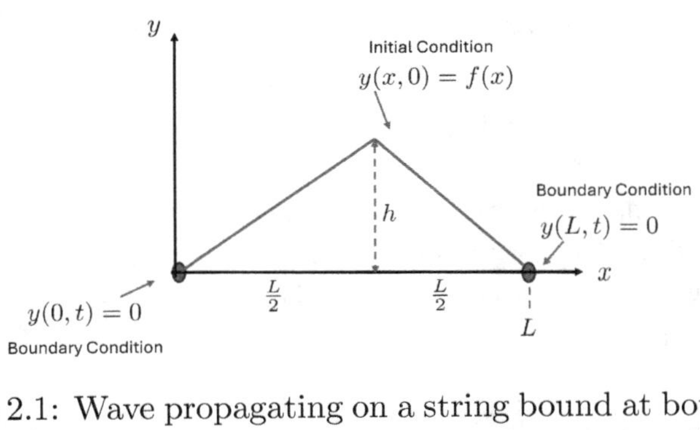

Figure 2.1: Wave propagating on a string bound at both ends Two boundary conditions and one initial condition shown.

We will begin by stating the wave equation for a mechanical wave traveling in a string. The partial differential equation (PDE) is :

$$\frac{\partial^2 y(x,t)}{\partial x^2} = \frac{1}{v^2}\frac{\partial^2 y(x,t)}{\partial t^2} \tag{2.1}$$

It represents a wave traveling in space and time with displacement in the y-direction. The $y(x,t)$ is the height of the wave at position, x, at any time, t. We need to know how $y(x,t)$ changes in space and time along the string.

Figure 2.1 shows the physical situation of the string being plucked at $t = 0$ by some force represented by $f(x)$. This is the initial condition. Additionally, we fix the two ends of the strings so they remain immobile. These are the two boundary conditions: $y(0,t) = y(L,t) = 0$. We will assume separation of variables to find the solution to the wave equation.

$$\begin{aligned} \text{Boundary Conditions: } y(0,t) &= 0 \\ y(L,t) &= 0 \\ \text{Initial Condition: } y(x,0) &= f(x) \end{aligned}$$

We will assume separation of variables for the PDE shown in equation (2.1) so we can turn the problem into a set of ordinary differential equations (ODE). The PDE for a wave traveling in a string is:

$$\frac{\partial^2 y(x,t)}{\partial x^2} = \frac{1}{v^2}\frac{\partial^2 y(x,t)}{\partial t^2} \tag{2.2}$$

Assuming the equation can be solved through separation of variables, the solution will be of the form

$$y(x,t) = F(x)G(t), \tag{2.3}$$

where $F(x)$ is a function of space (x) only and $G(t)$ is a function of time (t) only. Therefore, we have the following

$$\text{LHS: } \frac{\partial^2 y(x,t)}{\partial x^2} = \frac{\partial^2 F(x)}{\partial x^2} G(t)$$
$$\text{RHS: } \frac{1}{v^2}\frac{\partial^2 y(x,t)}{\partial t^2} = \frac{1}{v^2} F(x) \frac{\partial^2 G(t)}{\partial t^2}.$$

And when we set LHS=RHS we have

$$\frac{\partial^2 F(x)}{\partial x^2} G(t) = \frac{1}{v^2} F(x) \frac{\partial^2 G(t)}{\partial t^2}. \tag{2.4}$$

Now, we collect all terms that are functions of x one one side and all terms that are functions of t on the other side to obtain

$$\frac{\frac{\partial^2 F(x)}{\partial x^2}}{F(x)} = \frac{1}{v^2} \frac{\frac{\partial^2 G(t)}{\partial t^2}}{G(t)}.$$

This is the critical step in solving this PDE and it shows the usefulness of separation of variables. Since we are free to choose any values of x and any values of t, the only way the equation above could hold true is if both sides are equal to a constant. It will become apparent later, but for now let us choose the following constant for both LHS and RHS terms,

$$\frac{\frac{\partial^2 F(x)}{\partial x^2}}{F(x)} = \frac{1}{v^2} \frac{\frac{\partial^2 G(t)}{\partial t^2}}{G(t)} = -\omega^2. \tag{2.5}$$

Since each separate term is equal to a constant, we have just changed the problem of a PDE to a problem of an ODE such that

$$\frac{\frac{d^2F(x)}{dx^2}}{F(x)} = -\omega^2 \tag{2.6a}$$

$$\frac{1}{v^2}\frac{\frac{d^2G(t)}{dt^2}}{G(t)} = -\omega^2. \tag{2.6b}$$

ODE in space variable

Since we have simple boundary conditions in position, lets work with equation 2.6a first. We begin by multiplying both sides by $F(x)$ and re-state as a differential equation so that

$$\frac{d^2F(x)}{dx^2} = -\omega^2 F(x)$$

$$\frac{d^2F(x)}{dx^2} + \omega^2 F(x) = 0 \tag{2.7}$$

We have seen an equation such as equation 2.7 before. It is an ordinary second order homogeneous linear equation in space, undriven and undamped for which we already know its general solution as shown below.

$$F(x) = \alpha\cos(\omega x) + \beta\sin(\omega x) \tag{2.8}$$

Application of the boundary condition $y(0,t) = 0$ leads one to discard the $\cos(\omega x)$ term such that $\alpha = 0$ since $\cos(0) = 1$ and never zero.Therefore,

$$F(x) = \beta \sin(\omega x). \tag{2.9}$$

Now, we apply the second boundary condition $y(L,t) = 0$ to find that $F(x) = 0 = \beta \sin(\omega L)$, which begs the question, when is $\sin(\omega L) = 0$? From our prior knowledge, we know that $\sin(\omega L) = 0$ when $\omega L = \pi \to \omega = \frac{\pi}{L}$; $\omega L = 2\pi \to \omega = \frac{2\pi}{L}$ and so on. Thus, the boundary condition $\sin(\omega L) = 0$ will be satisfied when

$$\omega_n = \frac{n\pi}{L} \quad (n = 1, 2, 3, \cdots) \tag{2.10}$$

The solution, thus, becomes

$$F_n(x) = \beta_n \sin\left(\frac{n\pi}{L} x\right) \quad (n = 1, 2, 3, \cdots) \tag{2.11}$$

ODE in time variable

Now for the time variable. We go back to equation 2.6b and follow the same steps as before.

$$\frac{1}{v^2} \frac{\frac{d^2 G(t)}{dt^2}}{G(t)} = -\omega^2$$

$$\frac{d^2 G(t)}{dt^2} = -v^2 \omega^2 G(t)$$

Which gives,

$$\frac{d^2 G(t)}{dt^2} + (v\omega)^2 G(t) = 0. \tag{2.12}$$

As before, this is an ordinary second order homogeneous linear equation in time, undriven and undamped for which we already know its general solution as shown below:

$$G(t) = \gamma \cos(v\omega t) + \delta \sin(v\omega t) \tag{2.13}$$

After application of the initial condition $y(x,0) = f(x)$ we find that $\delta = 0$. Since $\sin(0) = 0$ there is no way the $\sin(v\omega t)$ term could result in $f(x)$ at $t = 0$. Thus, we obtain

$$G_n(t) = \gamma_n \cos\left(\frac{n\pi}{L} vt\right) \quad (n = 1, 2, 3, \cdots) \tag{2.14}$$

where we have used equation 2.10 for ω. γ_n will be determined by $f(x)$ in other words, how one plucks the string, since $y(x,0) = f(x) = \gamma$ at application of the initial condition. Now, it turns out that γ_n is given by the Fourier transform of $f(x)$ for each n. And each n represents a vibrational mode of the string. For example, $n = 1$ is the Fundamental Mode, $n = 2$ is the second harmonic, $n = 3$ is the third harmonic and so on. More on this later.

Full solution

We now write the full solution of $y(x,t)$ which is obtained by the assumed form of equation 2.3.

$$\begin{aligned} y_n(x,t) &= F(x)G(t) \\ &= \left[\beta_n \sin\left(\frac{n\pi}{L} x\right)\right] \left[\gamma_n \cos\left(\frac{n\pi}{L} vt\right)\right] \end{aligned}$$

Combining β_n with γ_n into A_n, which are just constant coefficients (real or complex) we obtain the final solution

$$y_n(x,t) = \sum_{n=1}^{\infty} A_n \sin\left(\frac{n\pi}{L}x\right)\cos\left(\frac{n\pi}{L}vt\right). \tag{2.15}$$

This is a linear combination of the solutions, which is a result of the differential equation being linear. In other words, if one solution satisfies the equation, a linear combination of solutions also satisfy the equation of motion.

It should be obvious that equation 2.15 describes a standing wave similar to the one introduced in class. The spatial component is separate from the temporal component. This fact can be more clearly seen if we work backwards from what we did in class by using the same trigonometric identity

$$\sin(C)\cos(D) = \frac{1}{2}\left[\sin(C+D) + \sin(C-D)\right],$$

where we are going to let $C \to \frac{n\pi}{L}x$ and $D \to \frac{n\pi}{L}vt$. Making those substitutions and using the trigonometric identity we get

$$y_n(x,t) = \sum_{n=1}^{\infty} \frac{A_n}{2}\left\{\sin\left[\frac{n\pi}{L}(x+vt)\right] + \sin\left[\frac{n\pi}{L}(x-vt)\right]\right\}. \tag{2.16}$$

You should be able to recognize this solution as a superposition of two traveling waves. The left sine term is a wave traveling to the left along the $-x$-direction. The sine term on the right is a wave traveling to the right along the $+x$-direction. These two waves travel along the string in opposite directions and give rise to a standing wave pattern (eqn 2.15) due to their superposition.

To gain further insight into equation 2.15, let's make the following substitutions in the sine and cosine terms:

$$\sin\left(\frac{n\pi}{L}\left(\frac{2}{2}\right)x\right) \rightarrow \sin\left(n\left(\frac{2\pi}{2L}\right)x\right).$$

But, we know that on a vibrating string, the wavelength of the fundamental mode is, $\lambda_1 = 2L$. Further, we know that the wavenumber, $k = \frac{2\pi}{\lambda}$. Therefore, $\frac{2\pi}{2L} \rightarrow k_1 = \frac{2\pi}{\lambda_1}$ and since $\lambda_n = \frac{\lambda_1}{n}$, more generally, we have $k_n = \frac{2\pi}{\lambda_n} = n\frac{2\pi}{\lambda_1}$. We can, thus, re-write the sine term of equation 2.15 as $\sin(k_n x)$ where $(n = 1, 2, 3, \cdots)$.

We make a similar substitution into the cosine term as follows:

$$\cos\left(\frac{n\pi}{L}\left(\frac{2}{2}\right)vt\right) \rightarrow \cos\left(2\pi\left(\frac{nv}{2L}\right)t\right).$$

But, we know that on a vibrating string, $f_1 = \frac{v}{2L}$, is the fundamental mode, where $v = \sqrt{T/\mu}$. Therefore we can re-write the cosine term of equation 2.15 as $\cos(2\pi n f_1 t)$ or $\cos(2\pi f_n t)$, where $f_n = nf_1 \quad (n = 1, 2, 3, \cdots)$.

Now we can re-write equation 2.15 with the substitutions made above so that we finally have

$$y_n(x,t) = \sum_{n=1}^{\infty} A_n \sin(k_n x)\cos(2\pi f_n t)$$

Or, since $\omega_n = 2\pi f_n$ this can be expressed in the more familiar form introduced in class:

$$y_n(x,t) = \sum_{n=1}^{\infty} A_n \sin(k_n x)\cos(\omega_n t). \tag{2.17}$$

In equation 2.17, each $y_n(x,t)$ is a normal mode with each A_n being the amplitude of each standing wave. This is an example of a Fourier Series. The spatial component, $\sin(k_n x)$, gives rise to the shape of the standing wave patterns such as: Fundamental mode ($n = 1$), second harmonic ($n = 2$), third harmonic ($n = 3$), etc. that we learned in class. The time dependence of each normal mode is the harmonic frequency given by $f_n = nf_1 = \frac{nv}{2L}$.

Fourier transform to find the coefficients, A_n.

The coefficients A_n in equation 2.17 can be found through the application of a Fourier Transform. Making use of the initial condition $y(x,0) = f(x)$, where $f(x)$ is some arbitrary function that describes how the string was initially plucked, we obtain

$$\begin{aligned} y_n(x,0) &= \sum_{n=1}^{\infty} A_n \sin(k_n x)\cos(0) = f(x) \\ &= \sum_{n=1}^{\infty} A_n \sin(k_n x) = f(x). \end{aligned}$$

The relationship

$$f(x) = \sum_{n=1}^{\infty} A_n \sin(k_n x) \tag{2.18}$$

states that the function which describes the plucking of the string can be approximated as a Fourier series. The coefficients, A_n can be calculated by computing the Fourier transform of $f(x)$,

$$A_n = \int_{-\infty}^{\infty} y_n^*(x) f(x) dx.$$

We will assume the wavefunction is normalized and further take advantage of the orthogonality of sine functions so that $y_n^*(x) = \left(\frac{2}{L}\right) \sin(k_n x)$. Our Fourier Transform will have the form

$$A_n = \frac{2}{L} \int_0^L \sin(k_n x) f(x) dx \tag{2.19}$$

for any general $f(x)$. Let us assume the string was plucked in the middle according to the following:

$$f(x) = \begin{cases} \frac{2h}{L} x & 0 \leq x \leq \frac{L}{2} \\ \frac{2h}{L}(L - x) & \frac{L}{2} \leq x \leq L. \end{cases}$$

Equation 2.19 will be

$$A_n = \frac{2}{L}\left[\int_0^{L/2} \sin\left(\frac{n\pi}{L}x\right)\frac{2h}{L}x\,dx\right] + \frac{2}{L}\left[\int_{L/2}^{L} \sin\left(\frac{n\pi}{L}x\right)\frac{2h}{L}(L-x)\,dx\right]. \tag{2.20}$$

After solving the integral, lots of algebra, and some cancellations the following is obtained as a result

$$A_n = \frac{8h}{n^2\pi^2}\sin\left(\frac{n\pi}{2}\right). \tag{2.21}$$

We now write down the entire equation that describes standing waves on a string for the specific way the sting was plucked in this case.

$$y_n(x,t) = \sum_{n=1}^{\infty}\left[\frac{8h}{n^2\pi^2}\sin\left(\frac{n\pi}{2}\right)\right]\sin(k_n x)\cos(\omega_n t) \tag{2.22}$$

$$y_n(x,t) = \sum_{n=1}^{\infty}\left[\frac{8h}{n^2\pi^2}\sin\left(\frac{n\pi}{2}\right)\right]\sin\left(\frac{n\pi}{L}x\right)\cos\left(\frac{n\pi}{L}vt\right) \tag{2.23}$$

Matlab program of equation above follows on next page.

Matlab Program to simulate:

$$f(x,t) = \left(\frac{2}{n^2 \pi^2} \sin\left(\frac{n\pi}{2}\right) \right) \cdot \sin\left(\frac{n\pi x}{L}\right) \cdot \cos\left(\frac{n\pi v t}{L}\right)$$

CODE from ChatGPT

```
% Parameters
L = 1;              % Length of the domain
v = 1;              % Wave speed
x = linspace(0, L, 1000);   % Spatial grid
t_vals = linspace(0, 5, 500);  % Time values to animate
max_n = 20;          % Number of modes to include in the sum

% Create figure
figure;
h = plot(x, zeros(size(x)), 'LineWidth', 2);
xlabel('x');
ylabel('f(x,t)');
title('Time Evolution of f(x,t)');
axis([0 L -0.5 0.5]);
grid on;

% Animation loop
for t = t_vals
    f_xt = zeros(size(x)); % initialize sum

    for n = 1:max_n
        coeff = (1 / (n^2 * pi^2)) * sin(n * pi / 2);
        f_xt = f_xt + coeff * sin(n * pi * x / L) .* cos(n * pi * v * t / L);
    end

    set(h, 'YData', f_xt);
    title(sprintf('Time Evolution of f(x,t), t = %.2f', t));
    drawnow;
end
```

Chapter 3

Classical Mechanics: Electromagnetic Standing Waves on Perfect Reflectors

Standing electromagnetic (EM) waves can form when light is incident onto perfect reflectors such as mirrors, in the case of laser cavities, or perfect conductors, in the case of microwave ovens. In either case, incident EM waves are reflected from the surface making for perfect reflectors. We will assume plane EM traveling waves of the form,

$$\vec{E}_y(x,t) = E_o \cos(kx - \omega t)\hat{j} \tag{3.1}$$

$$\vec{B}_z(x,t) = B_o \cos(kx - \omega t)\hat{k}, \tag{3.2}$$

incident onto perfect reflectors, where $B_o = E_o/c$. These obey Maxwell's equations for light traveling in vacuum:

$$\nabla \cdot \boldsymbol{E} = 0 \qquad \text{(Gauss's Law for Electric Field)} \tag{3.3}$$

$$\nabla \cdot \boldsymbol{B} = 0 \qquad \text{(Gauss's Law for Magnetic Field)} \tag{3.4}$$

$$\nabla \times \boldsymbol{E} = -\frac{\partial \boldsymbol{B}}{\partial t} \qquad \text{(Faraday's Law of Induction)} \tag{3.5}$$

$$\nabla \times \boldsymbol{B} = \frac{1}{c^2}\frac{\partial \boldsymbol{E}}{\partial t} \qquad \text{(Ampere-Maxwell's Law).} \tag{3.6}$$

Typical derivation of EM standing waves

The most common method to describe standing waves that form between perfect conductors is to start with the superposition of incident and reflected EM waves in order that the electric field at the surface of the perfect conductor is zero, which is required by nature. We will thus have the following

$$\vec{E}_i(x,t) = E_o \cos(kx - \omega t)\hat{\jmath} \qquad \text{(incident electric field wave)} \tag{3.7a}$$

$$\vec{E}_r(x,t) = -E_o \cos(-kx - \omega t)\hat{\jmath} \qquad \text{(reflected electric field wave).} \tag{3.7b}$$

And for the magnetic field we have

$$\vec{B}_i(x,t) = B_o \cos(kx - \omega t)\hat{k} \qquad \text{(incident magnetic field wave)} \tag{3.8a}$$

$$\vec{B}_r(x,t) = B_o \cos(-kx - \omega t)\hat{k} \qquad \text{(reflected magnetic field wave).} \tag{3.8b}$$

The superposition of these give

$$\vec{E}_y(x,t) = \vec{E}_i + \vec{E}_r = 2E_o \sin(kx)\sin(\omega t) \tag{3.9}$$

$$\vec{B}_z(x,t) = \vec{B}_i + \vec{B}_r = 2B_o \cos(kx)\cos(\omega t). \tag{3.10}$$

Clearly, equations (3.9) and (3.10) describe standing wave

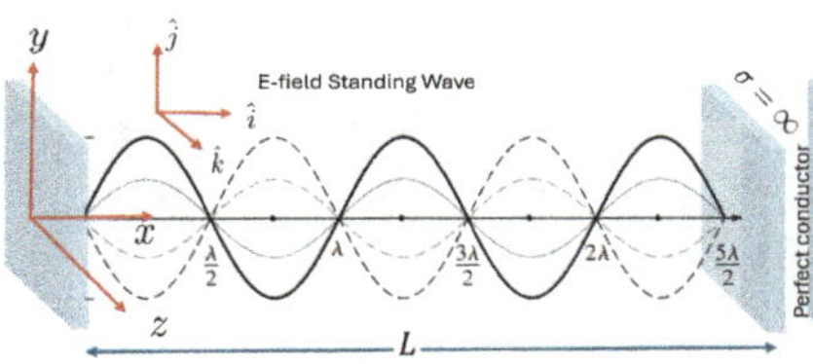

Figure 3.1: Standing wave of electric field.

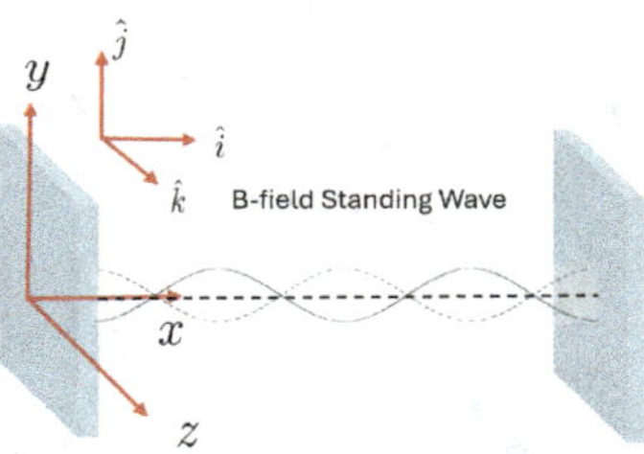

Figure 3.2: Standing wave of magnetic field.

since their spatial component is independent of the temporal component. Interestingly, the electic field at the surface of the perfect conductor is zero while the magnetic field is at its maximum amplitude, $2B_o$, at the surface. In fact, the electric field has nodal planes at $\sin(kx) = 0$. Since, $k = \frac{2\pi}{\lambda}$, these occur when $x = 0$; $\frac{2\pi}{\lambda}x = \pi \rightarrow x = \cancel{\pi}\frac{\lambda}{2\cancel{\pi}} = \frac{\lambda}{2}$; $\frac{2\pi}{\lambda}x = 2\pi \rightarrow x = \cancel{2\pi}\frac{\lambda}{\cancel{2\pi}} = \lambda$;$\frac{2\pi}{\lambda}x = 3\pi \rightarrow x = 3\cancel{\pi}\frac{\lambda}{2\cancel{\pi}} = 3\frac{\lambda}{2}$; and so on. In general, nodal planes occur at

$$x = n\left(\frac{\lambda}{2}\right) \quad (n = 0, 1, 2, 3, \cdots)$$

and anti-nodal planes occur at

$$x = (2n+1)\left(\frac{\lambda}{4}\right) \quad (n = 0, 1, 2, 3, \cdots)$$

as shown in Figure (3.1). As equation (3.10) and Figure (3.2) show, nodal planes for the magnetic field standing waves occur at locations of the antinodal planes in the electric field

$$x = (2n+1)\left(\frac{\lambda}{4}\right) \quad (n = 0, 1, 2, 3, \cdots)$$

and antinodal planes of magnetic field occur at nodal planes of the electric field,

$$x = n\left(\frac{\lambda}{2}\right) \quad (n = 0, 1, 2, 3, \cdots).$$

In other words, the electric and magnetic fields are out of phase by $\frac{\lambda}{4} \rightarrow \frac{\pi}{2} \rightarrow 90°$! Figures (3.3) and (3.4) visually show this on their respective planes of oscillation.

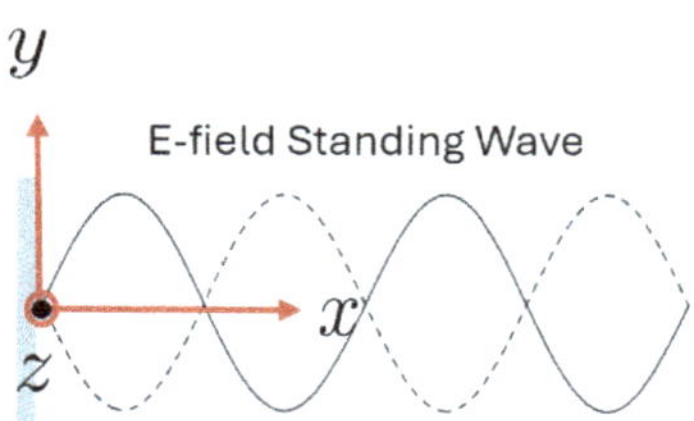

Figure 3.3: Standing wave of electric field on y-x plane.

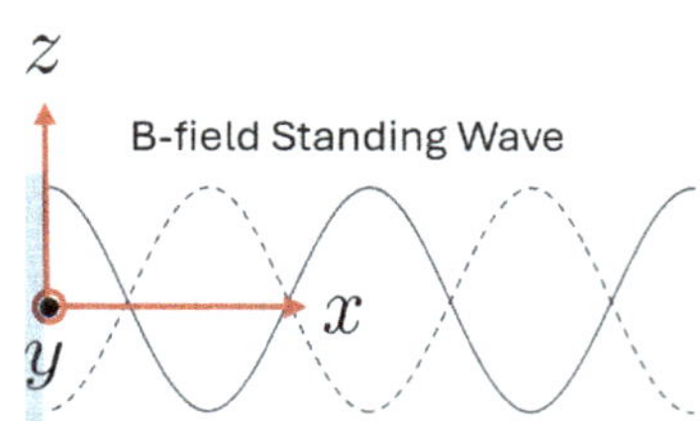

Figure 3.4: Standing wave of magnetic field on z-x plane.

Standing waves from decoupled wave equations

Now that we have described the typical method of deriving standing waves present in a reflecting cavity, we arrive at similar equations using equations of motion in the form of PDEs.

Through inspection of equations (3.5) and (3.6), one can see that the electric field and magnetic fields are coupled, that is the magnetic field depends on the electric field and vice versa. Decoupling them will lead to equations of motion for both the E-field and B-field. Additionally, it will allow us to use separation of variables to solve the PDEs. We start with Faraday's Law of Induction, equation (3.5), by taking the curl of both sides, using vector calculus and its properties, so that

$$\begin{aligned} \vec{\nabla} \times (\vec{\nabla} \times \vec{E}) &= \vec{\nabla} \times \left(-\frac{\partial \vec{B}}{\partial t} \right) \\ &= -\frac{\partial}{\partial t} \left(\vec{\nabla} \times \vec{B} \right) \\ &= -\frac{\partial}{\partial t} \left(\frac{1}{c^2} \frac{\partial \vec{E}}{\partial t} \right) \quad \text{Ampere-Maxwell's Law} \\ &= -\frac{1}{c^2} \frac{\partial^2 \vec{E}}{\partial t^2}. \end{aligned}$$

Now that we have the right hand side, we start on the left hand side with the identity

$$\vec{\nabla} \times (\vec{\nabla} \times \vec{E}) = \vec{\nabla}(\vec{\nabla} \cdot \vec{E}) - \nabla^2 \vec{E}.$$

But, Gauss Law for Electric Field shows that $\vec{\nabla} \cdot \vec{E} = 0$, which leads to

$$\nabla^2 \vec{E} = \frac{1}{c^2} \frac{\partial^2 \vec{E}}{\partial t^2}$$

Since we are assuming plane EM waves, $E_x = E_z = 0$, and only the y-component of electric field, E_y, is present, so that $\vec{E}(x,t) = E_y(x,t)\hat{j}$. Therefore, we get our first equation of motion in the form of a PDE, which is the wave equation for electric field:

$$\frac{\partial^2 E_y(x,t)}{\partial x^2} = \frac{1}{c^2}\frac{\partial^2 E_y(x,t)}{\partial t^2}. \tag{3.11}$$

The wave equation for magnetic field is of exactly the same form, and will only be presented here as:

$$\frac{\partial^2 B_z(x,t)}{\partial x^2} = \frac{1}{c^2}\frac{\partial^2 B_z(x,t)}{\partial t^2}. \tag{3.12}$$

Electric Field

Now that we have the equations of motion, we must work with both of them independently. We begin with the electric field and use the method of separation of variables as with waves on a string, then apply boundary and initial conditions. First, assume $E_y(x,t) = A(x)G(t)$ and substitute into equation (3.11), then divide both sides by $E_y(x,t) = A(x)G(t)$ to finally obtain

$$\frac{\frac{\partial^2 A(x)}{\partial x^2}}{A(x)} = \frac{1}{c^2}\frac{\frac{\partial^2 G(t)}{\partial t^2}}{G(t)} = -k^2,$$

where we define the wavenumber, k, as the constant to which they are both equal.

We follow the procedure as before to obtain

$$\frac{d^2A(x)}{dx^2} = -k^2A(x) \tag{3.13a}$$

$$\frac{1}{c^2}\frac{d^2G(t)}{dt^2} = -k^2G(t), \tag{3.13b}$$

which finally give us the ODE equations of motion

$$\frac{d^2A(x)}{dx^2} + k^2A(x) = 0 \tag{3.14a}$$

$$\frac{d^2G(t)}{dt^2} + \omega^2G(t) = 0, \tag{3.14b}$$

where we have used $\omega = ck$, the product of the speed of light, c, and wavenumber, k.

ODE spatial variable

The general solution to equation (3.14a) will be as before,

$$A(x) = \alpha\cos(kx) + \beta\sin(kx). \tag{3.15}$$

In this case, the two boundary conditions will be as follows: $E_y(0,t) = E_y(L,t) = 0$. The electric field at the surfaces of the reflectors will be zero. Therefore, application of the boundary condition at the reflector at $x = 0$ gives $A(0) = \alpha\cos(0) + \beta\sin(0) = 0 \to \alpha = 0 \longrightarrow A(x) = \beta\sin(kx)$. The second boundary condition at $E_y(L,t) = 0$ yields, $A(L) = \beta\sin(kL) = 0$. Since, $\beta = 0$ is a trivial solution, we find that $\sin(kL) = 0$ yields a family of solutions as before such that $k_n = \frac{n\pi}{L}$ $(n = 1, 2, 3, \cdots)$. In other words, only certain wavelengths, λ, can be sustained within the cavity length, L. These are $\lambda_n = \frac{2L}{n}$ $(n = 1, 2, 3, \cdots)$, which means only the following frequencies within the length of the cavity can appear: $f_n = \frac{c}{\lambda_n}$, where c is the speed of llight. The solution in space can, thus, be written as

$$A_n(x) = \beta_n \sin(k_n x) \quad (n = 1, 2, 3, \cdots). \tag{3.16}$$

ODE time variable

We now turn our attention to the time dependent ODE, equation (3.14b). As before, the general solution is

$$G(t) = \gamma \cos(\omega t) + \delta \sin(\omega t). \tag{3.17}$$

In this case, we will have to turn to the acceleration of the electric field at $t = 0$, which is zero since the electric field oscillates as a standing wave. The initial condition thus becomes:

$$\left.\frac{\partial^2 E_y(x,t)}{\partial t^2}\right|_{t=0} = 0.$$

We take the second time derivative of equation (3.17).

$$\frac{dG(t)}{dt} = -\omega\gamma \sin(\omega t) + \omega\delta \cos(\omega t)$$
$$\frac{d^2G(t)}{dt^2} = -\omega^2\gamma \cos(\omega t) - \omega^2\delta \sin(\omega t)$$

Now we set

$$\frac{d^2G(0)}{dt^2} = -\omega^2\gamma \cancelto{1}{\cos(0)} - \cancelto{0}{\omega^2\delta \sin(0)} = 0 \longrightarrow \gamma = 0.$$

Therefore, the solution in time can be written as

$$G(t) = \delta \sin(\omega t). \tag{3.18}$$

Since $\omega = ck$, for our k_n we have that $\omega_n = ck_n$. The full solution of $E_y(x,t) = A(x)G(t)$ is then

$$E_y(x,t) = A_n \sin(k_n x) \sin(\omega_n t) \quad (n = 1, 2, 3, \cdots), \tag{3.19}$$

where we have combined the coefficients for β_n and δ_n into A_n. Comparison of equations (3.19) and (3.9) can be seen to be of the same standing wave form, a product of sine waves in space and time.

Magnetic Field

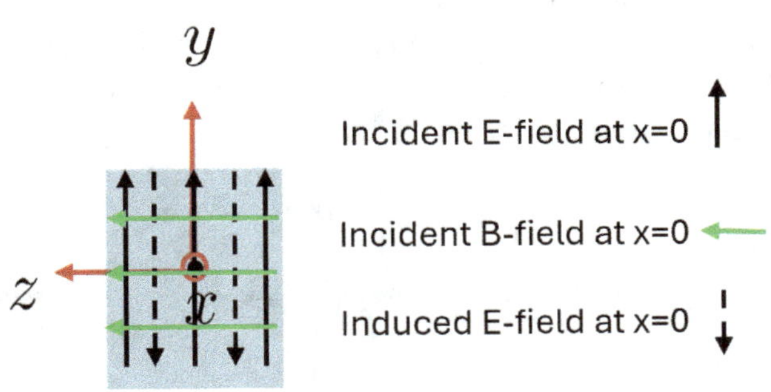

Figure 3.5: Electric field and Magnetic field oscillations on surface of perfect conductor at x=0.

As stated previously, at $x = 0$ the magnetic field is at its maximum value, $|2B_o|$, on the surface of a perfect conductor while the electric field is zero due to an induced electric field at the surface which cancels the effect of the incoming electric field as shown in Figure (3.5), thus setting up a standing wave. The magnetic field varies sinusoidally from $+2B_o$ to $-2B_o$ on the surface of the conductor, $x = 0$, as shown in Figure (3.4). Consequently, the slope of the variation in magnetic field with respect to x is zero since the variation is at the surface of the conductor. Additionally, for all x at $t = 0$, the magnetic field is stationary. In other words, there is no propagation of the magnetic field in time.

We, thus, begin the usual procedure with equation (3.12) by assuming separation of variables and perfom operations similar to that done with the electric field. We will assume a solution of the form, $B_z(x,t) = F(x)G(t)$ which we then substitute into equation (3.12), divide both sides by the assumed solution. The ratio of variables in space will equal the ratio of the variables in time which can only occur if they are both equal to a constant that we will define as the wavenumber, k. We then present, the result as

$$\frac{d^2F(x)}{dx^2} + k^2F(x) = 0 \tag{3.20a}$$

$$\frac{d^2G(t)}{dt^2} + \omega^2G(t) = 0, \tag{3.20b}$$

where we have used $\omega = ck$, the product of the speed of light, c, and wavenumber, k.

ODE spatial variable

The general solution to equation (3.20a) will be as before,

$$F(x) = \alpha \cos(kx) + \beta \sin(kx). \tag{3.21}$$

However, in this case, we will apply the boundary conditions as mentioned above.

$$\left.\frac{\partial B_z(x,t)}{\partial x}\right|_{x=0} = 0,$$

which results in the following:

$$\begin{aligned}\frac{dF(x)}{dx} &= -\alpha k \sin(kx) + \beta k \cos(kx)\\ \frac{dF(0)}{dx} &= \cancelto{0}{-\alpha k \sin(0)} + \beta k \cancelto{1}{\cos(0)} = 0.\end{aligned}$$

This leads to $\beta = 0$. We will apply the next boundary condition $B_z(0,t) = B_z(L,t) = |2B_o|$ to $F(x) = \alpha \cos(kx)$ now, which results in $F(0) = \alpha \cos(0) = |2B_o| \longrightarrow \alpha = |2B_o|$. Our function in space now looks like

$$F(x) = |2B_o| \cos(kx) \tag{3.22}$$

At $x = L$, the spatial function becomes $F(L) = |2B_o| \cos(kL) = |2B_o|$. This boundary condition provides no additional information and its application will overconstrain our function. However, it is interesting to note that the only way this boundary condition can happen is when $\cos(kL) = |\pm 1|$ which is describing the maximum amplitude oscillating between $+2B_o$ and $-2B_o$ on the surface of the conductor as stated before. The allowed k values turn out to be integer multiple of π so that $k_n = \frac{n\pi}{L}$ $(n = 0, 1, 2, 3, \cdots)$. Interestingly, we can also find the nodal planes described by $F(x)$ which have to occur when $\cos(kx) = 0$. These can happen when $kx = \frac{\pi}{2} \to \frac{2\pi}{\lambda}x = \frac{\pi}{2} \to x = \frac{\cancel{\pi}}{2}\frac{\lambda}{2\cancel{\pi}} \to x = \frac{\lambda}{4}$. In fact, for any odd integer multiples of $\lambda/4$ there will be nodal planes at positions given by $x = \frac{\lambda}{4}(2n+1)$ $(n = 0, 1, 2, 3, \cdots)$.

ODE time variable

We now look at equation (3.20b) to resolve the time functionality of the equation of motion. Here the general solution will be

$$G(t) = \gamma \cos(\omega t) + \delta \sin(\omega t) \tag{3.23}$$

Here we will apply the initial conditions mentioned above.

$$\left.\frac{\partial B_z(x,t)}{\partial t}\right|_{t=0} = 0,$$

which results in the following:

$$\begin{aligned}\frac{dG(t)}{dt} &= -\gamma\omega \sin(\omega t) + \delta\omega \cos(\omega t)\\ \frac{dG(0)}{dt} &= \cancelto{0}{-\gamma\omega \sin(0)} + \delta\omega \cancelto{1}{\cos(0)} = 0.\end{aligned}$$

This leads to $\delta = 0$. Therefore,

$$G(t) = \gamma \cos(\omega t). \tag{3.24}$$

We now state the full solution $B_z(x,t) = F(x)G(t)$ as follows $B_z(x,t) = \gamma|2B_o|\cos(kx)\cos(\omega t)$. Now, we began by assuming incident and reflecting plane waves. To satisfy Maxwell's equations, $B_o = E_o/c$. From equation (3.19), we find that we can replace $\gamma|2B_o| \to A_n/c$, where we have lumped in the constant coefficents into A_n. In addition, there are certain allowed wavevectors, k_n as well as frequencies, $f_n \to \omega_n$. Therefore, our final form of the magnetic field solution is

$$B_z(x,t) = \frac{A_n}{c}\cos(k_n x)\cos(\omega_n t) \quad (n = 1, 2, 3, \cdots). \tag{3.25}$$

This looks similar in form to equation (3.10).

Here, we restate the electric field and magnetic field equations as follows

$$E_y(x,t) = A_n \sin(k_n x)\sin(\omega_n t) \quad (n = 1, 2, 3, \cdots)$$

$$B_z(x,t) = \frac{A_n}{c}\cos(k_n x)\cos(\omega_n t) \quad (n = 1, 2, 3, \cdots)$$

Chapter 4

Quantum Mechanics: Probability Standing Waves- Particle in a Box

We consider a particle (not shown) confined within walls of infinite potential. In between the walls, the potential is zero, $U(x) = 0$ so the particle moves freely back and forth within the walls due to repeated reflections. The distance between the walls is, L and the particle moves along x only. That is, in one dimension.

We present the time-dependent Schrödinger's equation in

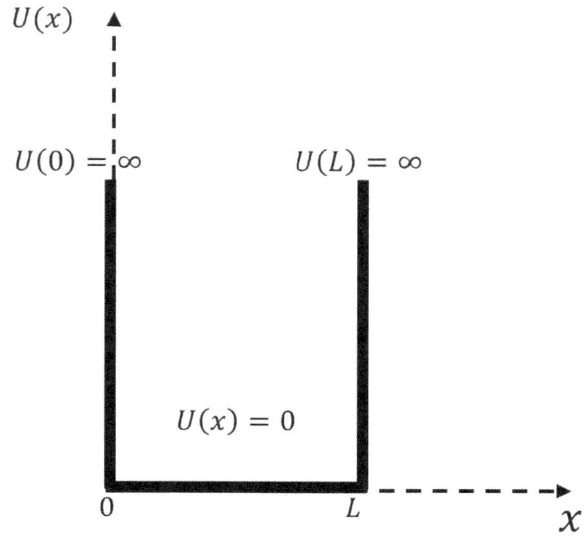

Figure 4.1: Free particle bounded by infinite potentials separated by a length, L. Two boundary conditions.

one dimension as follows:

$$i\hbar\frac{\partial}{\partial t}\Psi(x,t) = \left[-\frac{\hbar^2}{2m}\frac{\partial^2}{\partial x^2} + U(x,t)\right]\Psi(x,t) \tag{4.1}$$

Here, we will consider the case where $U(x,t)$ is time independent and further $U(x) = 0$ in the region $0 < x < L$ so equation (4.1) becomes

$$i\hbar\frac{\partial}{\partial t}\Psi(x,t) = \left[-\frac{\hbar^2}{2m}\frac{\partial^2}{\partial x^2}\right]\Psi(x,t). \tag{4.2}$$

As before, the equation will have solutions that can be solved through separation of variables of the form

$$\Psi(x,t) = \psi(x)g(t) \tag{4.3}$$

where $\psi(x)$ is a function of space (x) only and $g(t)$ is a function of time (t) only. Therefore, we have the following

$$\text{LHS: } i\hbar\frac{\partial\Psi(x,t)}{\partial t} = i\hbar\frac{\partial g(t)}{\partial t}\psi(x)$$

$$\text{RHS: } \left[-\frac{\hbar^2}{2m}\frac{\partial^2}{\partial x^2}\right]\Psi(x,t) = \left[-\frac{\hbar^2}{2m}\frac{\partial^2\psi(x)}{\partial x^2}\right]g(t).$$

And when we set LHS=RHS we have

$$i\hbar\frac{\partial g(t)}{\partial t}\psi(x) = \left[-\frac{\hbar^2}{2m}\frac{\partial^2\psi(x)}{\partial x^2}\right]g(t). \tag{4.4}$$

Now, we collect all terms that are functions of t one one side and all terms that are functions of x on the other side to obtain

$$\frac{i\hbar\frac{\partial g(t)}{\partial t}}{g(t)} = \frac{\left[-\frac{\hbar^2}{2m}\frac{\partial^2\psi(x)}{\partial x^2}\right]}{\psi(x)} = E$$

As before, the only way the equation above could hold true is if both sides are equal to a constant, which we have called E. Thus we obtain two ODEs as follows:

$$-\frac{\hbar^2}{2m}\frac{\partial^2\psi(x)}{\partial x^2} = E\psi(x) \tag{4.5a}$$

$$i\hbar\frac{\partial g(t)}{\partial t} = Eg(t). \tag{4.5b}$$

Time-independent Schrödinger Equation

Equation (4.5a) is a function of x only and is known as the time-independent Schrödinger equation. We write it in the form of the well-known second ODE:

$$\frac{\partial^2 \psi(x)}{\partial x^2} + \frac{2mE}{\hbar^2}\psi(x) = 0. \tag{4.6}$$

First, we define a wavenumber as

$$k = \sqrt{\frac{2mE}{\hbar^2}} \tag{4.7}$$

and re-write it as

$$\frac{\partial^2 \psi(x)}{\partial x^2} + k^2\psi(x) = 0. \tag{4.8}$$

Now we can clearly see that this looks like an ordinary differential equation we have solved before. It's solution is a superposition of two oscillating functions as before:

$$\psi(x) = \alpha\cos(kx) + \beta\sin(kx) \tag{4.9}$$

Application of the boundary conditions $U(0) = U(L) = \infty$ means that the particle is confined to the region $(0, L)$. The probability that the particle exists outside this region is zero. This leads to the requirement that $\psi(0) = \psi(L) = 0$, which are the two boundary conditions. We, thus, find that after application of the $\psi(0) = 0$ boundary condition $\alpha = 0$ since $\cos(0) = 1$. Therefore,

$$\psi(x) = \beta\sin(kx). \tag{4.10}$$

Now, we apply the second boundary condition $\psi(L) = 0$ to find that $\psi(L) = 0 = \beta \sin(kL)$. The trivial solution of $\beta = 0$ yields no information. However, we now ask the question, when is $\sin(kL) = 0$? We know that $\sin(kL) = 0$ when $kL = \pi \rightarrow k = \frac{\pi}{L}$; $kL = 2\pi \rightarrow k = \frac{2\pi}{L}$ and so on. Thus, the boundary condition $\sin(kL) = 0$ will be satisfied when

$$k_n = \frac{n\pi}{L} \quad (n = 1, 2, 3, \cdots) \tag{4.11}$$

The solution, thus, becomes

$$\psi_n(x) = \beta \sin(k_n x) = \beta \sin\left(\frac{n\pi}{L}x\right) \quad (n = 1, 2, 3, \cdots) \tag{4.12}$$

In the Copenhagen interpretation, $|\psi(x)|^2$, represents the probability of finding the particle within the region $(0, L)$. Therefore, we can find the coefficient β by normalizing the wavefunction so that the probability of finding the particle within the region $(0, L)$ is equal to one such that

$$\int_0^L \psi_n^*(x)\psi_n(x)dx = 1.$$

This means that

$$\int_0^L \beta^2 \sin^2(k_n x)dx = \beta^2 \int_0^L \sin^2(k_n x)dx = \beta^2\left(\frac{L}{2}\right) = 1$$

$$\Longrightarrow \beta = \sqrt{\frac{2}{L}}.$$

Therefore, the normalized equation (4.12) becomes

$$\psi_n(x) = \sqrt{\frac{2}{L}} \sin\left(\frac{n\pi}{L}x\right) \quad (n = 1, 2, 3, \cdots) \tag{4.13}$$

Before we go on to solve for the time-dependent ODE, equation (4.5b), we first take a closer look at our result for the wavenumber, equation (4.11). We have previously defined our wavenumber according to equation (4.7). We, thus, set equation (4.11) equal to equation (4.7) and solve for E so that

$$\sqrt{\frac{2mE}{\hbar^2}} = \frac{n\pi}{L} \longrightarrow \frac{2mE}{\hbar^2} = \left(\frac{n\pi}{L}\right)^2 \longrightarrow E_n = \frac{\hbar^2}{2m}\left(\frac{n\pi}{L}\right)^2.$$

This states that the allowed energy of the freely moving particle within this box can only take on certain discrete energy levels described by the quantum number (n) according to:

$$E_n = \left(\frac{\hbar^2\pi^2}{2mL^2}\right) n^2 \quad (n = 1, 2, 3, \cdots). \tag{4.14}$$

Since $\hbar = \frac{h}{2\pi}$, we can also express (4.14) as

$$E_n = \left(\frac{h^2}{8mL^2}\right) n^2 \quad (n = 1, 2, 3, \cdots). \tag{4.15}$$

With a minor amount of algebra, the energy difference between adjacent energy levels is found to be

$$\Delta E_n = E_{n+1} - E_n = \left(\frac{h^2}{8mL^2}\right)(2n + 1) \quad (n = 1, 2, 3, \cdots). \tag{4.16}$$

Time-dependent ODE

We now turn out attention to the time dependent portion of $\Psi(x,t)$ equation (4.5b), which we re-state here with some minor re-arrangement,

$$\begin{aligned} i\hbar\frac{\partial g(t)}{\partial t} &= Eg(t) \\ \frac{dg(t)}{dt} &= \left(\frac{i}{i}\right)\frac{1}{i\hbar}Eg(t) \\ \frac{dg(t)}{dt} &= \frac{i}{i^2\hbar}Eg(t) \\ \frac{dg(t)}{dt} &= -\frac{i}{\hbar}Eg(t). \end{aligned}$$

This is easily integrable to obtain

$$g(t) = e^{\left(-\frac{iE}{\hbar}t\right)}.$$

However, we must take into account that E depends on n. So we have

$$g_n(t) = e^{\left(-\frac{iE_n}{\hbar}t\right)}. \tag{4.17}$$

We first write down the stationary state solution for equation (4.2)

$$\Psi_n(x,t) = \psi_n(x)g_n(t)$$

$$\Psi_n(x,t) = \psi_n(x)e^{\left(-\frac{iE_n}{\hbar}t\right)} \quad (n = 1, 2, 3, \cdots) \tag{4.18}$$

Equation (4.18) is called a stationary state solution because the probability density does not depend on time, although the spatial component of the wave function moves in time.

This can be clearly seen upon calculation of the probability density

$$
\begin{aligned}
|\Psi_n(x,t)|^2 &= \int_0^L \Psi_n^*(x,t)\Psi_n(x,t)dx \\
&= \int_0^L \psi_n^*(x)\cancel{e^{\left(\frac{iE_n}{\hbar}t\right)}}\psi_n(x)\cancel{e^{\left(-\frac{iE_n}{\hbar}t\right)}}dx \\
&= \int_0^L \psi_n^*(x)\psi_n(x)dx,
\end{aligned}
$$

which does not depend on time.

Now we write down the entire solution for the time-dependent Schrödinger equation for the special case of a particle in a box in one dimension, where we have combined the constant for $\psi_n(x)$ with the constant for $g_n(t)$ into general constants, c_n, to obtain a superposition of equation (4.18).

$$
\Psi_n(x,t) = \sum_{n=1}^{\infty} c_n \sin\left(\frac{n\pi}{L}x\right) e^{\left(-\frac{iE_n}{\hbar}t\right)} \quad (n = 1, 2, 3, \cdots) \tag{4.19}
$$

These sets of solutions are not stationary states as they will depend on time.

Bibliography

[1] Georgi, H, The Physics of Waves, New Jersey, Prentice-Hall, 1993.

[2] McQuarrie, D.A., Quantum Chemistry, California, University Science Books, 1983.

[3] Zettili, N., Quantum Mechanics: Concepts and Applications, West Susex, UK, John Wiley and Sons, Ltd., 2009.

[4] Young, H.D., Freedman, R.A.,University Physics, 15th Ed., USA, Pearson Education, Inc., 2012.

www.ingramcontent.com/pod-product-compliance
Lightning Source LLC
LaVergne TN
LVHW011513200926
848484LV00001B/31